Fennec Fox

Julie Murray

Abdo Kids Junior
is an Imprint of Abdo Kids
abdobooks.com

Abdo
MINI ANIMALS
Kids

abdobooks.com

Published by Abdo Kids, a division of ABDO, P.O. Box 398166, Minneapolis, Minnesota 55439.

Abdo Kids Junior™ is a trademark and logo of Abdo Kids.

Printed in the United States of America, North Mankato, Minnesota.

102019

012020

THIS BOOK CONTAINS RECYCLED MATERIALS

Photo Credits: Alamy, iStock, Shutterstock

Production Contributors: Teddy Borth, Jennie Forsberg, Grace Hansen

Design Contributors: Christina Doffing, Candice Keimig, Dorothy Toth

Library of Congress Control Number: 2019941263

Publisher's Cataloging-in-Publication Data

Names: Murray, Julie, author.

Title: Fennec fox / by Julie Murray

Description: Minneapolis, Minnesota : Abdo Kids, 2020 | Series: Mini animals | Includes online resources and index.

Identifiers: ISBN 9781532188800 (lib. bdg.) | ISBN 9781644943021 (pbk.) | ISBN 9781532189296 (ebook) | ISBN 9781098200275 (Read-to-Me ebook)

Subjects: LCSH: Fennec fox--Juvenile literature. | Foxes--Juvenile literature. | Desert animals--Juvenile literature. | Animal size--Juvenile literature. | Size and shape--Juvenile literature.

Classification: DDC 599.775--dc23

Table of Contents

Fennec Fox

The fennec fox is small.

It is the smallest fox!

It is 8 inches (20 cm) tall. It weighs 2 to 3 pounds (0.9 to 1.4 kg).

It lives in northern Africa.

Africa

It makes its home in the sand.

It **burrows** in the **dunes**.

It has fur. The fur is sandy in color.

It has big ears. They are 6 inches (15 cm) long!

It has a long tail.

The tip is black.

It has fury feet. The sand is hot. The fur protects the feet!

It eats rodents and insects.

It also eats plants.

Let's Compare!

fennec fox

Height: 8 inches (20 cm)
Weight: 3 pounds (1.4 kg)
Ear Length: 6 inches (15 cm)

red fox

Height: 17 inches (43 cm)
Weight: 6–31 pounds (3–14 kg)
Ear Length: 4 inches (10 cm)

Glossary

burrow
to dig into the earth.

dune
a mound or hill of sand built up by wind.

Index